The Christmas Truce

when peace broke out on the Western Front

based on true events

*written and illustrated
by Duane Porter*

ISBN 978-1-962937-01-6

Illustrations and cover by Duane Porter

Published by
Buried Treasure Publishing

Zephyrhills, Florida

BuriedTreasurePublishing.com

Printed in the U.S.A.

The Christmas Truce is based on true events that occurred around the Christmas season in 1914-1915. For a brief period, the guns fell silent and opposing forces forged an unofficial truce between them.

The characters and illustrations serve to bring this story to life. This message of peace, despite all odds, is one that needs to be shared.

I find it comforting that this story unfolded during Christmas. The thought that this season fills us with hope, that the message from the angel of "Peace on Earth" can still be relevant today, binds our desires with those of the soldiers embroiled in the conflict of the Great War.

This is what stories are for. They touch us, they stir a passion for something we yearn for. In the end, they drive us to seek better paths for the future.

Duane Porter

December, 1914. It couldn't have been a more dismal Christmas Eve.

Here I was, standing guard in the freezing cold. Yesterday's drizzle only made the mud in the bottom of the trench worse. Bloody Huns. This war was all their fault. I'd like to pop one of them when he sticks his head up.

Lights moved in the distance. I peered through the dusk at the German line. Definitely lights moving above the parapet.

I squeezed off a round across no-man's land, hoping to hit something. A chorus of shouts came from the enemy line, only 60 yards away. "No shoot! No shoot! Tannenbaum! Ist Tannenbaum! No shoot tonight, Jock!"

"Tannen-bomb?" I muttered. I stared through the gathering gloom and saw that the lights were candles attached to small pine trees. "What kind of bomb is a Tannen-bomb?"

"Christmas tree, mate." My buddy, George, squelched through the muck to crouch beside me. "They're celebrating Christmas, German style."

"They're bloody crazy."

"Nah, they're just thinking of home like us, I reckon. Don't you miss Scotland, Ian? It's the first Christmas of this stinking war for all of us."

I nodded. "Maybe. Listen..."

Voices raised in song floated across from the enemy line. The tune was familiar, but the words were strange: *"Stille Nacht, heilige Nacht, alles schlaft, einsam wacht..."*

"They're singing *Silent Night*," I muttered. "The Huns are singing Christmas carols."

We stood in the filthy trench listening to the Germans sing two or three more songs. It was surreal hearing the music here, out in a field in Belgium, so far away from home.

"Here now, we can't let them have all the fun!" George exclaimed. He broke into a loud rendition of *Good King Wenceslas.* He clapped me on the shoulder and waved his hand for me to join in. I grinned and added my baritone while George sloshed down the line waking others to help out.

We traded carols with the Germans for several hours, until an unusual quiet drifted over the front.

I braved a final look over the top. A long row of Tannenbaum flickered warmly along the German trench line, glowing with the light of a hundred candles.

A lone voice echoed merrily across the gap. "No shoot tonight, Jock!" I slumped back down into the trench and tried to get some sleep.

The next morning it dawned cold but sunny. I forced down some tasteless rations and wondered what today would bring. I didn't have to wait long.

"No shoot! No shoot!" voices called from the enemy lines. Several Germans climbed out of their trench and stood along the parapet, unarmed!

"What are they doing?" I whispered to myself. I raised my rifle to fire, then paused. "They're unarmed. They asked us not to shoot."

Slowly I lowered my gun. "It's Christmas, for God's sake."

More unarmed Germans joined the growing crowd strolling along the deadly parapet. No one fired. Finally, curiosity got the best of me. I set my gun carefully in a dry spot and clambered wearily over the top.

"No shoot!" I called out, surprised at how husky my voice sounded.

A group of Germans saw me and waved for me to come closer. "Any of you lads speak English?" I asked as I drew near.

"Ya, I speak a little," an impossibly young-looking soldier responded with a grin. "Merry Christmas!"

"Merry Christmas!" I replied dreamily. Was I still asleep?

"Hey, Ian, what's going on?" George joined me, also weaponless. "Who's your friend?"

"Oh, this is..."

"Hans." The German rescued me with a smile and a wink. "And you are...?"

"George. You've met Ian, here. Merry Christmas! Hey, what's that?"

A hare scampered across the field, darting back and forth. Germans and Scots alike tried to catch it, both hungry for a fresh meal.

The Germans eventually captured the hare but the friendly competition of the chase broke the ice.

We began trading all kinds of things between ourselves and our enemies: Jam for sausage, chocolate for cigars, patches for buttons. We found butterscotch and chocolates in our tin "Princess Mary" boxes to barter with gifts from King George.

The Germans, not to be outdone, received their own presents from the Kaiser: a large meerschaum pipe and tobacco for each enlisted man, while their officers got fine cigars.

Soon our battalion chaplain asked a German officer if we could use this truce to bury the dead from both sides that had fallen in no-man's land, the deadly killing zone between the trenches. They agreed, and we worked side by side to dig great mass graves, one for each army.

After the burial, we held a service for the dead. The Germans held their spiked helmets in their hands, and we Scots stood reverently, wearing our kilts soiled with the mud from the trenches. The chaplain read the 23rd Psalm. Hans translated and read it in German, too.

We stood together with our fallen comrades.

"Well, I'm not in any hurry to get back to the war," I said as the service ended.

"Nein," Hans agreed. "What do you do for fun?"

George produced a brown leather ball from behind his back. "We like to beat the Kaiser's team in football!"

"Where did you get that?" I laughed.

"Doesn't matter! You're on!" Hans chuckled and raced off to gather a team.

It was a spirited game, and a close one at that, with the Germans edging out the 6th Battalion of the Gordon Highlanders three to two.

We built small fires in no-man's land, and sat around sharing wine and cognac until we retired back to our trenches for the night.

Several days passed like this. We were careful when higher ranking commanders would visit the front lines; they had learned that there was something going on. Something they didn't like. They wanted us to fight, to kill.

When they came, we jumped back into our trenches and fired shots well over the heads of our friends on the opposite side.

Then came the moment I dreaded.

I was on watch, and a German strolled on his parapet unconcerned.

"Private!"

I turned to find a general behind me. I thought they were noisier than that.

"Is that a German out there?"

"Sir, yes, sir!"

"Well, shoot him!"

Now he decides to be noisy. I wheeled and fired a shot over the German's head. He continued his walk, unperturbed.

"Is he moving around too much for you, soldier?"

"No, sir!" I cried and fired a shot just to the German's side. He looked around, seeking where the sound of the bullet came from.

"Private..."

I fired once again, close enough this time that the startled soldier turned and dove head first into his trench.

"Hmph." The general moved on down the trench line.

Within a few days, changes came to the front. Some units were exchanged out for others, artillery bombardment resumed in earnest, and the fighting began again. The military leaders on both sides wanted war, not peace. For now, anyway.

I'll remember that Christmas, though, and the fragile truce that followed from Christmas Eve of 1914 into the pale days of January. I hope that the whole world will remember.

www.ingramcontent.com/pod-product-compliance
Lightning Source LLC
LaVergne TN
LVHW072135070426
835513LV00003B/111